Immortal DNA: Living Forever as an Eternal Spirit

1. How Long is DNA Helix?...10

2. Big Bang to End of Time...20

3. Speck of Time Lifetime...30

4. Finding your Purpose...40

5. On Fire for Jesus...50

6. Living in the Spirit...61

7. Your Pending Creations...73

8. Making Ripples in Time...83

9. Becoming an Eternal Star...94

10. Legacy of Immortal DNA...105

Immortal DNA: Living Forever as an Eternal Spirit takes readers on an extraordinary journey into the spiritual and eternal dimensions of existence. This transformative book explores the divine blueprint encoded within your DNA, connecting your physical reality with the eternal spirit that transcends time. With ten captivating chapters, it unveils profound insights into the mysteries of creation, the brevity of earthly life, and the timeless nature of the soul.

Discover how the structure of the DNA helix mirrors infinite potential, linking your spiritual essence to the Creator's eternal plan. From the Big Bang to the end of time, you'll learn to see the universe as a divine construct where your soul plays a vital role. Explore how even the smallest moments in life ripple through eternity, shaping a legacy of love, purpose, and faith.

Whether you're seeking your divine purpose, aligning with the Holy Spirit, or creating meaningful spiritual impacts, this book offers wisdom and guidance. Dive deep into the transformative power of living a Christ-centered life and embrace the concept of becoming an eternal star, shining your unique light across generations.

Immortal DNA: Living Forever as an Eternal Spirit is more than a book—it's a spiritual awakening, a call to recognize your role as a co-creator with God, and a reminder of your eternal significance. If you're ready to unlock the secrets of eternal living, this book will illuminate your path to a life of purpose, faith, and everlasting impact.

About Scott Perdue

Scott Perdue is a dynamic entrepreneur, author, and community leader with a life rooted in faith, family, and service. A devoted Christian, Scott has been married for over 20 years and is the proud father of four children—two girls and two boys. His passion for personal development and spiritual growth is reflected in his prolific writing career, having authored over 100 books, most of which focus on self-help and Christian themes. His books have touched the lives of countless readers seeking guidance on how to lead a fulfilling, faith-centered life.

For over 15 years, Scott has been a dedicated member of GUTS Church, a place he fondly refers to as "It Takes GUTS to Serve the Lord." His service to the church and community extends beyond attendance; he spent six years as a representative for the GUTS Food Bank, where he managed the movement of wholesale goods to help those in need. Scott also led a successful Maximized Manhood study group based on Edwin Cole's teachings, further exemplifying his commitment to fostering spiritual growth among men.

An accomplished entrepreneur, Scott has started and operated over 30 businesses, ranging from pest control to contracting. He is the founder of Universal Bug Man, a pest control service where Scott earned a reputation as a "pest control superhero." His entrepreneurial ventures include Tulsa Furniture Wholesale, Tulsa Auction Spot, and Builderhaus Unlimited, among others. Scott's business acumen extends to the health and wellness industry, where his company HCG Medical helped over 20,000 clients lose weight, generating over $6.5 million in sales in its best year.

Scott Perdue is a man of many talents, driven by his faith and dedication to serving others through his varied enterprises and writing.

Follow Scott Perdue on YouTube, Facebook & Visit UniversalWholesaleStore.com

Published Books by Scott Perdue (Buy Today on Amazon)

Christian Books by Scott Perdue:

Biblical Entrepreneur Leadership: Amplified Leverage Business Skills Book & Workbook

Biblical Men's Leadership Skills: Becoming an Amplified Christian Superstar Book & Workbook

Unleashing Biblical Manhood: Taking Ground Like a Warrior Book & Workbook

Promised Land Leadership: Leading an Army Like Joshua

Wilderness Wisdom of Moses: Timeless Life-Changing Leadership Lessons

Rules of Christianity According to Paul Book & Workbook

Provisional Miracles of Jesus: Provision through Supernatural Means Book & Workbook

Kingdom Money: Unlocking Biblical Secrets to Financial Success

The King's Highway: Lean into Jesus for Accelerated Success

Walk in the Works of the Lord: An Amplified Passion Understanding

God's River: Getting into the Kingdom Family Flow

Forgiven & Unoffendable: The Power of Walking Righteously

God is Real: Knowing the Spirit - A Journey Through Faith, Miracles, and Divine Presence

The Gift of Light: A Journey of Spiritual Growth for Life Expansion

On Fire For Jesus: Bring Plasma Energy to Your Heart Pump

Immortal DNA: Living Forever as an Eternal Spirit

Earth is God's Beach Ball: Celebrate the Legacy of Joyful Living

Faith in the Wilderness: Biblical Lessons for Strength and Spiritual Growth

Living on Purpose: A Comprehensive Guide to a Meaningful and Fulfilling Life

Praying for Others: Unlocking your God-Given Authority to Change Lives

Speaking in Tongues: Snippets of Life Improvement Code

Be Fruitful and Multiply: A Biblical Guide to Family Planning and Takes

Biblical Map of the Garden of Eden: Where does this Mysterious Garden Exist?

Methuselah: The Biblical Legacy of Noah's Grandfather

Love's Crossroads: The Rewards of Suffering for Love

Features of a Great Christian Camp: A Priority Spiritual Foundation

Daily Mercy: A Journey Through God's Grace Every Morning

Self Help Books by Scott Perdue:

You Are the Masterpiece: Center of the Universe Life Experience

Legacy Blueprint: How to Build a Generational Legacy

Accomplishing Greatness: 10 Legendary Skill Sets of Self-Made Millionaires

The Passive Income Playbook: 10 Game Changing Strategies to Build Wealth

Beginners Guide to Investing in the Future: Gain Wealth from Cutting Edge Sectors

Motivation for Creation: Unlocking the Spark Within

Master Productivity: Unlock your Path to Success

10 Step Productivity Plan: A Guide to Increasing Life's Results

Mindset of Productivity: A Defined Focused Journey

Mindful Love: Embracing Self Love Through Mindfulness and Compassion

Mindfulness for Personal Growth: Transform Your Life One Moment at a Time

The Ultimate Guide to Winning Friends and Influencing People: Master Communication

The Human Connection: Unlocking the Secrets to Understanding and Relating to Others

Stress Free Living: Simple Strategies for Modern Life

Mind Switch: Are you Over-Thinking Negative Thoughts?

Mastering Self-Control: Unleashing the Power of Discipline for Success in Every Aspect of Life

Rising From The Ashes: How to Rebuild When Life Falls Apart

Unlocking Secrets to Weight Loss: A Comprehensive Guide to Science, Nutrition, and Wellness

Effective Diet Supplements for Weight Loss

The Body Detox Blueprint: 10 Essential Steps to Cleanse, Heal, and Revitalize Your Body

Secret 1000 Calorie Cryogenic Diet

Learn to Enjoy Reading: Your Ultimate Guide to Loving Books

The Ultimate Blueprint to Comedy: Your Guide to Mastering Humor and Making People Laugh

Decluttering Your Home: Take Control of Your Space, One Step at a Time

Real Estate Needs Observation: Hot to Bring Light to Entropy & Chaos

Business Books by Scott Perdue:

Legendary Business Skills: How to Think like an Entrepreneur

Seal the Deal: Mastering Sales Objections to Close Every Sale

10 Step Marketing Launch: Ultimate Guide for a Business Advertising Start Up

Email Marketing Success: 10 Ways to Master Business Email Advertising Strategy

Controlled Decent: How to Close a Business

How to Start a Business Networking Group: Learn to Organize and Motivate Business Leaders

Negotiate Like an Auctioneer: Mastering the Art of Persuasion and Control

Auction House Blueprint: How to Win Bids and Host a Successful Auction

How to Run an Antique Shop: Restoring Antique Relics to Modern Living

Secondhand Success: A Complete Guide to Running a Profitable Used Furniture Store

The Thrift Store Playbook: How to Build, Manage, and Thrive in the Resale Business

How to Start an RV Park: Your Roadmap to Success

Turn Rapids to Revenue: How to Run a Profitable River Float Business

Science Books by Scott Perdue:

Creation of Your Galactic Record: Big Bang, DNA, Creation of the Universe. Boom!

Symphony of Life: How Human DNA Plays Like Music

Quantum Cosmos: The Wave Function of the Universe

An Astronauts Heavenly Perspective: Planet, Society and Economy

Earth is the Seed of Life: A Geometric Flower of Life

Infinite Plants in Every Seed

Dodecahedron Earth: Exploring the Geometric Key to the Flower of Life

Pangaea Cracked Open: A Pre-Flood World without Oceans

Ancient Cathedral Architecture: A Language Of Semantics Lost in Time

Power Independence: DIY Guide to Building Off-Grid Energy Systems

Harvesting Heaven: The Ultimate Guide to DIY Rainwater Collection Systems

Farming Tactics for the Sahara Desert: Ultimate Gardening Guide for Arid Takes

Easy to Find Herbal Remedies

How to Build Free Energy Lighting: 10 Effective Easy to Build Free Energy Lights

Dynamic Forces: Exploring the Undeniable Power of Movement

Creative Books by Scott Perdue:

Zeppelin Airship Enterprise: The Future of Flight and Travel Reimagined

Ancient Plasma Energy Weapons Revealed: The Lost Technology of Energy Weapons

Echoes of Camelot: Unveiling the Secrets and Legends of the Knights

Secret Treasures of Rome Revealed: Explore the Ancient Architecture of Rome

The Egyptian Ankh: Secrets of Eternal Life and Ancient Wisdom

Giants, Nephilim, and the Legacy of Humanity: From Ancient Myths to Modern Mysteries

Prophecy of the Seven Suns: Exploring Parhelia in Biblical Prophecy

Epic Scavenger Hunt of Machu Picchu

Adventures of Buying an Island: Edge of Your Seat Suspense Thriller Adventure

My Neighbor is an Inventor: A Journey into Wilson's World of Innovation

Adventures of the Zoo Janitor: Growing Responsibility By Excellence

Exile's Genesis: Chronicles of the New Frontier

Relics of Oklahoma: Route 66 Treasure Hunt

The Oklahoma Waterfall Hunt

How Long is DNA Helix?

The concept of Immortal DNA reveals the divine blueprint embedded within every person, not just as a physical reality but as a profound spiritual truth. Human DNA, the molecular code that defines biological life, offers incredible insight into the eternal nature of the spirit. Through understanding the intricate design of DNA, we uncover parallels between our physical existence and our spiritual identity. This journey illuminates how to live eternally in the Spirit, unlocking the truth about our immortal nature through the lens of scripture and divine revelation.

How Long Physically is a DNA Helix?

DNA, or deoxyribonucleic acid, is a microscopic structure that encodes the instructions for life. Incredibly, if the DNA from a single cell were unraveled and stretched out, it would measure about two meters in length. When multiplied across the trillions of cells in the human body, the total DNA could stretch to the sun and back several times. This immense length, compressed into the nucleus of a cell, is a testament to the infinite potential encoded within every human being.

The physical length of DNA is awe-inspiring, yet it is only a fraction of its significance. DNA acts as the master instruction manual for every function of the body. Similarly, our spiritual DNA holds the eternal blueprint of God's purpose for us. Just as DNA provides continuity of life through replication, our spiritual essence connects us to eternity.

Psalm 139:13-14 (AMP) captures this beautifully:
"For You formed my innermost parts; You knit me [together] in my mother's womb. I will give thanks and praise to You, for I am fearfully and wonderfully made; Wonderful are Your works, And my soul knows it very well."

This scripture highlights the divine craftsmanship behind both our physical and spiritual existence. DNA, though finite in its earthly role, points to an infinite Creator whose design transcends the physical realm.

The Infinite Potential Encoded Within Human DNA

DNA is not merely a molecule—it is a reservoir of potential. It encodes the instructions to build and sustain life, but it also symbolizes the unlimited potential God has placed within every individual. This potential is not limited to earthly achievements but extends into eternal significance.

Ephesians 2:10 (AMP) affirms this:
"For we are His workmanship [His own master work, a work of art], created in Christ Jesus [reborn from above—spiritually transformed, renewed, ready to be used] for good works, which God prepared [for us]

beforehand [taking paths which He set], so that we would walk in them [living the good life which He prearranged and made ready for us]."

The infinite potential encoded in human DNA mirrors our spiritual capacity to fulfill God's purpose. Just as DNA can adapt, repair, and evolve, our spirit can grow, transform, and align with God's will. This dynamic potential is activated through faith and a relationship with the Holy Spirit.

Spiritual Parallels of DNA's Structure and Continuity

The double-helix structure of DNA is not only a scientific marvel but also a spiritual metaphor. It reflects the dual nature of humanity: body and spirit, intertwined and inseparable. The continuous replication of DNA across generations symbolizes the eternal nature of the spirit. While the body is temporary, the spirit carries on, connecting us to God's eternal plan.

Ecclesiastes 3:11 (AMP) speaks to this eternal design:

"He has made everything beautiful and appropriate in its time. He has also planted eternity [a sense of divine purpose] in the human heart [a mysterious longing which nothing under the sun can satisfy, except God]—yet man cannot find out (comprehend, grasp) what God has done [His overall plan] from the beginning to the end."

The continuity of DNA ensures that life persists, while the continuity of the spirit ensures that our connection to God remains unbroken. This unending connection is the foundation of immortality in the Spirit.

The Connection Between Physical and Spiritual Immortality

DNA points us to a deeper reality: that while the body may perish, the spirit endures. This connection is made clear

through Christ's promise of eternal life. John 3:16 (AMP) declares:
"For God so [greatly] loved and dearly prized the world, that He [even] gave His [One and] only begotten Son, so that whoever believes and trusts in Him [as Savior] shall not perish, but have eternal life."

Physical DNA is a temporary vessel for the eternal essence within us. The resurrection of Jesus Christ demonstrates the ultimate transformation from physical mortality to spiritual immortality. In Him, we see the fulfillment of the promise that death is not the end but a transition to eternal life.

1 Corinthians 15:42-44 (AMP) explains this transformation:
"So also is the resurrection of the dead. The body that is sown is perishable and decays, but the body that is resurrected is imperishable [and immortal]. It is sown in dishonor; it is raised in glory. It is sown in weakness; it is raised in strength. It is sown

a natural body [mortal, suited to earth]; it is raised a spiritual body [immortal, suited to heaven]."

Living in Spirit: Unlocking the Truth About Immortal DNA

To live in the Spirit is to embrace the truth that your spiritual DNA has been encoded with eternity. This requires a fundamental shift in perspective, from seeing life as finite to understanding it as eternal.
Romans 8:14-16 (AMP) explains this transformation:
"For all who are allowing themselves to be led by the Spirit of God are sons of God. For you have not received a spirit of slavery leading again to fear [of God's judgment], but you have received the Spirit of adoption as sons [the Spirit producing sonship] by which we [joyfully] cry, 'Abba! Father!' The Spirit Himself testifies and confirms together with our spirit [assuring us] that we [believers] are children of God."

Living in the Spirit involves aligning your life with the divine purpose God has written into your spiritual DNA. This includes:

1. Faith in Christ

Faith activates the immortality of your spirit. By believing in Jesus, you receive the gift of eternal life and awaken to your divine purpose.

2. Walking in Daily Surrender

Living in the Spirit requires daily surrender to God's guidance. Galatians 5:16 (AMP) encourages:

"But I say, walk habitually in the Holy Spirit [seek Him and be responsive to His guidance], and then you will certainly not carry out the desire of the sinful nature."

3. Bearing Eternal Fruit

A Spirit-filled life produces fruit that has eternal impact, such as love, joy, peace, and kindness (Galatians 5:22-23 AMP).

The Immortal DNA Code: How to Live Eternally in the Spirit

The Immortal DNA Code offers a framework for unlocking and living out your eternal identity:

1. D – Discover Your Divine Design
Understand that you are uniquely created for a purpose. Meditate on Jeremiah 29:11 (AMP):
"'For I know the plans and thoughts that I have for you,' says the Lord, 'plans for peace and well-being and not for disaster, to give you a future and a hope.'"

2. N – Nurture Your Spiritual Nature
Commit to growing in your relationship with God through prayer, worship, and study of the Word. 2 Timothy 3:16-17 (AMP) emphasizes the importance of Scripture:
"All Scripture is God-breathed [given by divine inspiration] and is profitable for instruction, for conviction, for correction, and for training in righteousness."

3. A – Activate Your Eternal Calling
Step boldly into the purpose God has set for your life. Ephesians 2:10 (AMP) reminds us:
"For we are His workmanship [His own master work], created in Christ Jesus for

good works, which God prepared beforehand."

Conclusion

Understanding Immortal DNA is the key to realizing your eternal nature. Through Christ, you are not just a temporary being but a spiritual creation destined for eternity. By aligning your life with the Immortal DNA Code, you unlock the truth about living forever as an eternal spirit.

Revelation 21:6-7 (AMP) affirms this promise:
"Then He said to me, 'It is done. I am the Alpha and the Omega, the Beginning and the End. To the one who thirsts I will give [water] from the fountain of the water of life without cost. He who overcomes will inherit these things; and I will be his God, and he will be My son.'"

Embrace the truth of your immortal DNA and live a life that echoes into eternity.

Big Bang to End of Time

The concept of Immortal DNA connects humanity's physical and spiritual existence to a divine continuum, spanning from the moment of creation to eternity. In this narrative, the energy unleashed at the Big Bang is more than a scientific event; it is the spark of divine power that continues to resonate within every person's DNA. By tracing humanity's spiritual journey, understanding time as a divine construct, and unlocking the truth about living in the Spirit, we uncover the blueprint for eternal life.

From Big Bang to the End of Time: A Divine Perspective

The Big Bang, often described as the origin of the universe, marks the moment when God's creative energy was unleashed into the cosmos. Genesis 1:1-3 (AMP) captures the essence of this act:

"In the beginning God (Elohim) created [by forming from nothing] the heavens and the earth. The earth was formless and void, and darkness was upon the face of the deep. The Spirit of God was moving (hovering, brooding) over the face of the waters. And God said, 'Let there be light'; and there was light."

This light, both physical and spiritual, set into motion the unfolding of time and space. From that moment, humanity's journey began, not just as physical beings but as carriers of God's eternal essence. The energy of creation, encoded in the fabric of the universe, is also imprinted within human DNA.

Tracing Humanity's Spiritual Journey

Humanity's spiritual journey is a continuum, stretching from the divine spark of creation to the promise of eternal life. This journey transcends the boundaries of birth and death, as life is not confined to a single

earthly existence but extends into eternity. Ecclesiastes 3:11 (AMP) affirms:
"He has made everything beautiful and appropriate in its time. He has also planted eternity [a sense of divine purpose] in the human heart [a mysterious longing which nothing under the sun can satisfy, except God]—yet man cannot find out [comprehend] what God has done [His overall plan] from the beginning to the end."

This eternal purpose is encoded in our spiritual DNA. From Adam, created in God's image, to the redemption offered through Christ, humanity's story is one of reconnection with the divine. Romans 5:18-19 (AMP) explains this reconciliation:
"So then as through one trespass [Adam's sin] there resulted condemnation for all men, even so through one act of righteousness there resulted justification of life to all men. For just as through one man's disobedience his many were made sinners, so through the obedience of the

One [Jesus Christ] the many will be made righteous."

This spiritual lineage continues unbroken for those who align with God's eternal plan.

Time as a Divine Construct

Time, as experienced by humans, is a construct designed to give order to life. However, God exists outside of time, seeing the beginning and the end simultaneously. Isaiah 46:9-10 (AMP) declares:
"Remember [carefully] the former things which I did from ages past; For I am God, and there is no one else; I am God, and there is no one like Me, Declaring the end and the result from the beginning, And from ancient times the things which have not [yet] been done, Saying, 'My purpose will be established, And I will do all that pleases Me and fulfills My purpose.'"

While humans perceive time as linear, from birth to death, God's perspective is eternal.

The Big Bang, often thought of as the beginning of time, is merely the starting point of the physical universe. It serves as a reminder that God's creative power transcends temporal limitations.

Human DNA, passed down through generations, mirrors this divine continuity. The replication and preservation of genetic information symbolize the eternal spirit that resides within us, unbound by the confines of time.

Connecting the Big Bang's Energy to the Eternal Spirit

The energy released during the Big Bang echoes within each person as the divine spark of life. This energy, encoded in human DNA, carries the potential for both physical and spiritual immortality. While science reveals the intricate mechanisms of DNA, faith unveils its deeper significance.

John 1:1-4 (AMP) ties creation to Christ, the source of eternal life:

"In the beginning [before all time] was the Word (Christ), and the Word was with God, and the Word was God Himself. He was [continually existing] in the beginning [co-eternally] with God. All things were made and came into existence through Him; and without Him not even one thing was made that has come into being. In Him was life [and the power to bestow life], and the life was the Light of men."

This passage reveals that the energy of creation is Christ Himself, the eternal Word. His light illuminates the eternal spirit within us, making immortality possible through Him.

Living in Spirit: Unlocking the Truth About Immortal DNA

To live in the Spirit is to embrace the eternal nature encoded in your DNA. This requires shifting your focus from the temporary to the eternal, recognizing that your physical body is a vessel for an everlasting spirit.

Galatians 5:25 (AMP) challenges us to live in alignment with the Spirit:
"If we claim to live by the [Holy] Spirit, we must also walk by the Spirit [with personal integrity, godly character, and moral courage—our conduct empowered by the Holy Spirit]."

Living in the Spirit involves:
    1. Recognizing Your Divine Identity
You are more than your physical body; you are a spiritual being created in God's image. Genesis 1:27 (AMP) declares:
"So God created man in His own image, in the image and likeness of God He created him; male and female He created them."
    2. Cultivating a Relationship with God
Eternal life begins with knowing God intimately. John 17:3 (AMP) states:
"Now this is eternal life: that they may know You [the only true supreme and

sovereign] God, and [in the same manner know] Jesus [as the] Christ whom You have sent."

3. Living with Eternal Purpose
Every action has eternal significance. By aligning your life with God's purpose, you fulfill the destiny encoded in your spiritual DNA.

The Immortal DNA Code: How to Live Eternally in the Spirit

The Immortal DNA Code provides a practical framework for unlocking and living out your eternal identity:

1. D – Discover Your Eternal Origin
Recognize that the energy of creation resides within you. Meditate on Colossians 1:16 (AMP):
"For by Him all things were created in heaven and on earth, things visible and invisible, whether thrones or dominions or rulers or authorities; all things were created and exist through Him [that is, by His activity] and for Him."

2. N – Nurture Your Connection to the Spirit

Cultivate a life led by the Holy Spirit through prayer, worship, and study of the Word. Romans 8:14 (AMP) emphasizes:

"For all who are allowing themselves to be led by the Spirit of God are sons of God."

3. A – Align Your Life with Eternity

Shift your focus from temporary pursuits to eternal significance. Matthew 6:19-20 (AMP) instructs:

"Do not store up for yourselves treasures on earth, where moth and rust destroy, and where thieves break in and steal. But store up for yourselves treasures in heaven, where neither moth nor rust destroys, and where thieves do not break in and steal."

Conclusion

The truth of Immortal DNA reveals that you are a spiritual being, created with the energy of eternity. From the Big Bang to the end of time, God's plan has been to connect you to His eternal purpose. By embracing the Immortal DNA Code, you unlock the key to living forever in the Spirit.

Revelation 22:13 (AMP) encapsulates this eternal promise:
"I am the Alpha and the Omega, the First and the Last, the Beginning and the End [the Eternal One]."

Embrace your divine identity and live a life that echoes into eternity, guided by the Spirit and anchored in the truth of your immortal DNA.

Speck of Time Lifetime

The human lifetime, when viewed through the lens of eternity, is but a fleeting moment—a mere speck of time in the vast expanse of God's eternal plan. Yet within this fleeting existence lies profound significance. This concept of Immortal DNA connects our temporal lives to an eternal purpose, offering insight into how our seemingly small actions can create everlasting ripples. Through this understanding, we unlock the truth about living in the Spirit and discover the code for eternal life.

The Speck of Time in the Context of Eternity

Psalm 90:4 (AMP) captures the essence of time's fleeting nature:
"For a thousand years in Your sight are like yesterday when it is past, or as a watch in the night."

Our earthly lives, often averaging only a handful of decades, are but a breath when measured against eternity. James 4:14 (AMP) echoes this truth:
"Yet you do not know [the least thing] about what may happen in your life tomorrow. What is secure in your life? You are merely a vapor [like a puff of smoke or a wisp of steam from a cooking pot] that is visible for a little while and then vanishes into thin air."

While life on earth is transient, it holds immense potential for eternal impact. Each moment, though fleeting, carries the possibility of aligning us with God's eternal purposes. This understanding shifts our perspective, compelling us to live intentionally and with the awareness that our actions reverberate through eternity.

The Eternal Impact of Small Moments

Every moment in life presents an opportunity to make choices that transcend time. Galatians 6:7-8 (AMP) reminds us of the eternal consequences of our actions: *"Do not be deceived, God is not mocked [He will not allow Himself to be ridiculed, nor treated with contempt, nor allow His precepts to be scornfully set aside]; for whatever a man sows, this and this only is what he will reap. For the one who sows to his flesh [his sinful capacity, his worldliness, his disgraceful impulses] will reap from the flesh ruin and destruction, but the one who sows to the Spirit will from the Spirit reap eternal life."*

Small acts of love, kindness, and obedience to God can have an enduring impact, shaping not only our eternal destiny but also influencing others. Consider the widow's offering in Mark 12:41-44 (AMP): *"For they all contributed from their surplus; but she, from her poverty, put in all she had [all she had to live on]."*

Though her act might have seemed insignificant in human terms, it was monumental in God's eyes. This story illustrates that even the smallest actions, when aligned with God's will, carry eternal weight.

Living with an Eternal Perspective

To live with an eternal perspective means to prioritize what truly matters—faith, love, and the fulfillment of God's purpose. Colossians 3:2 (AMP) instructs us:
"Set your mind and keep focused habitually on the things above [the heavenly things], not on things that are on the earth [which have only temporal value]."

When we view our lives through the lens of eternity, our values and priorities shift:
- We invest in relationships that glorify God.
- We use our resources to advance His kingdom.
- We cultivate spiritual growth over material gain.

Jesus emphasized the importance of living with eternity in mind in Matthew 6:19-20 (AMP):

"Do not store up for yourselves treasures on earth, where moth and rust destroy, and where thieves break in and steal. But store up for yourselves treasures in heaven, where neither moth nor rust destroys, and where thieves do not break in and steal."

This eternal mindset transforms how we approach even mundane tasks, imbuing them with purpose and significance.

The Truth About Immortal DNA

Human DNA carries the blueprint for life, encoding the instructions necessary for physical existence. But it also serves as a metaphor for our spiritual design. Just as DNA is a continuous chain that connects generations, so too does our spiritual essence link us to God's eternal plan.

Ephesians 2:10 (AMP) reveals the divine blueprint within us:

"For we are His workmanship [His own master work, a work of art], created in Christ Jesus [reborn from above—spiritually transformed, renewed, ready to be used] for good works, which God prepared [for us] beforehand [taking paths which He set], so that we would walk in them [living the good life which He prearranged and made ready for us]."

This verse highlights that we are intentionally designed with a purpose that transcends time. Our spiritual DNA contains the potential for eternal life, a gift unlocked through Christ.

Living in the Spirit: The Path to Eternity

Living in the Spirit means aligning our lives with the Holy Spirit, who empowers us to fulfill our divine purpose. Romans 8:14 (AMP) states:

"For all who are allowing themselves to be led by the Spirit of God are sons of God."

This spiritual alignment transforms how we live, enabling us to overcome the limitations of our earthly nature and embrace our eternal identity. It involves:

    1. Surrendering to God's Will

Proverbs 3:5-6 (AMP) reminds us to trust in God:

"Trust in and rely confidently on the Lord with all your heart, and do not rely on your own insight or understanding. In all your ways know and acknowledge and recognize Him, and He will make your paths straight and smooth [removing obstacles that block your way]."

    2. Walking in Love

Galatians 5:22-23 (AMP) lists the fruits of the Spirit:

"But the fruit of the Spirit [the result of His presence within us] is love [unselfish concern for others], joy, [inner] peace, patience [not the ability to wait, but how we

act while waiting], kindness, goodness, faithfulness, gentleness, self-control."
Living in the Spirit means embodying these qualities, which have eternal value.

3. Fulfilling God's Mission
Matthew 28:19-20 (AMP) commands us to make disciples:
"Go therefore and make disciples of all the nations [help the people to learn of Me, believe in Me, and obey My words], baptizing them in the name of the Father and of the Son and of the Holy Spirit, teaching them to observe everything that I have commanded you."

The Immortal DNA Code: How to Live Eternally in the Spirit

The Immortal DNA Code provides a framework for living a life of eternal significance:

1. D – Develop an Eternal Perspective
Recognize that your earthly life is a speck of time compared to eternity. Psalm 39:5 (AMP) states:

"Behold, You have made my days as short as hand widths, and my lifetime is as nothing in Your sight. Surely every man at his best is a mere breath [a wisp of smoke, a vapor that vanishes]! Selah."

2. N – Nurture Spiritual Growth
Spend time in prayer, worship, and studying God's Word to deepen your relationship with Him. 2 Peter 3:18 (AMP) encourages: *"But grow [spiritually mature] in the grace and knowledge of our Lord and Savior Jesus Christ. To Him [be] glory (honor, majesty, splendor), both now and to the day of eternity. Amen."*

3. A – Act with Eternal Purpose
Let every decision reflect your eternal values. Ephesians 5:15-16 (AMP) advises: *"Therefore see that you walk carefully [living life with honor, purpose, and courage; shunning those who tolerate and enable evil], not as the unwise, but as wise [sensible, intelligent, discerning people], making the very most of your time [on earth, recognizing and taking advantage of each opportunity and using it with wisdom*

and diligence], because the days are [filled with] evil."

Conclusion

Though earthly life is fleeting—a mere speck of time—it holds the key to eternity. By living in the Spirit and embracing the truths of Immortal DNA, we unlock the potential to live forever as eternal beings. Every moment, every decision, and every act of love contributes to a legacy that transcends time.

Revelation 22:5 (AMP) offers a glimpse of our eternal destiny:
"And there will no longer be night; they have no need for lamplight or sunlight, because the Lord God will illumine them; and they will reign as kings forever and ever [throughout the eternities of the eternities]."

Live today with eternity in mind, and let your life echo into forever.

Finding Your Purpose

Life's ultimate fulfillment comes from understanding and living out the purpose for which we were created. This purpose is not merely a product of chance or circumstance but is intricately woven into our very being—our Immortal DNA. This spiritual blueprint carries the essence of God's design for each of us, connecting us to eternity and guiding us toward a divine mission. By unlocking this purpose, we align our lives with eternal significance and discover the profound truth of living forever in the Spirit.

Unlocking the Divine Purpose Embedded in Your Soul

From the moment of creation, God placed within each of us a unique purpose, a divine calling that reflects His eternal plan. This purpose is not hidden but is embedded within the core of who we are. Psalm 139:16 (AMP) reveals this truth:

"Your eyes have seen my unformed substance; and in Your book were all written the days that were appointed for me, when as yet there was not one of them [even taking shape]."

Every aspect of our lives—our talents, passions, and even our challenges—points toward the divine purpose God has written for us. Discovering this purpose requires introspection, faith, and a willingness to surrender to God's will. It is a journey of uncovering the eternal significance of our existence and aligning ourselves with His higher plan.

How Spiritual DNA Guides Your Mission in Life

Just as physical DNA serves as a roadmap for our biological development, spiritual DNA acts as a guide for our eternal mission. This spiritual DNA carries the imprint of God's design, uniquely equipping each of us for a specific role in His kingdom.

1. Understanding the Call

Every person has been called by God to fulfill a unique mission. This calling is not confined to worldly accomplishments but is rooted in eternal significance. Romans 11:29 (AMP) states:

"For the gifts and the calling of God are irrevocable [for He does not withdraw what He has given, nor does He change His mind about those to whom He gives His grace or to whom He sends His call]."

This calling is embedded in our spiritual DNA, waiting to be activated through faith and obedience.

2. Discovering Your Unique Gifts

God has equipped each of us with spiritual gifts and talents to fulfill our purpose. 1 Peter 4:10 (AMP) encourages:

"Just as each one of you has received a special gift [a spiritual talent, an ability graciously given by God], employ it in serving one another as [is appropriate for] good stewards of God's multifaceted grace [faithfully using the diverse, varied gifts and

abilities granted to Christians by God's unmerited favor]."

These gifts are like keys, unlocking doors to opportunities that align with our divine mission.

3. Following the Spirit's Guidance
Walking in alignment with our spiritual DNA requires us to live in the Spirit, allowing the Holy Spirit to guide our steps. Galatians 5:16 (AMP) reminds us:

"But I say, walk habitually in the Holy Spirit [seek Him and be responsive to His guidance], and then you will certainly not carry out the desire of the sinful nature [which responds impulsively without regard for God and His precepts]."

By yielding to the Spirit, we stay on the path God has designed for us, fulfilling our eternal mission.

Aligning Personal Goals with Eternal Significance

One of the most profound aspects of understanding Immortal DNA is recognizing the eternal significance of our actions and choices. Our earthly goals and aspirations must align with God's purpose for our lives to bear eternal fruit. Matthew 6:33 (AMP) directs us:

"But first and most importantly seek (aim at, strive after) His kingdom and His righteousness [His way of doing and being right—the attitude and character of God], and all these things will be given to you also."

1. Shifting from Temporal to Eternal Goals

Many people pursue temporary achievements—wealth, status, and success—without considering their eternal impact. However, living in the Spirit calls us to focus on goals that glorify God and contribute to His kingdom. Colossians 3:2 (AMP) exhorts:

"Set your mind and keep focused habitually on the things above [the heavenly things], not on things that are on the earth [which have only temporal value]."

2. Using Talents for God's Glory

Our gifts and talents are not meant solely for personal gain but are tools for advancing God's purposes. 1 Corinthians 10:31 (AMP) teaches:

"So then, whether you eat or drink or whatever you do, do all to the glory of [our great] God."

Whether in our careers, relationships, or ministries, we must seek to glorify God through our efforts.

3. Building an Eternal Legacy

The choices we make today impact eternity, leaving a legacy that outlives us. Proverbs 13:22 (AMP) states:

"A good man leaves an inheritance to his children's children, and the wealth of the sinner is stored up for [the hands of] the righteous."

This legacy is not merely material but spiritual, as we pass down faith, love, and wisdom to future generations.

Living in the Spirit: Unlocking the Truth About Immortal DNA

Living in the Spirit is essential for activating the Immortal DNA within us. This lifestyle requires us to surrender completely to God, allowing His Spirit to lead us in fulfilling our divine purpose.

1. Surrendering to God's Will

True fulfillment comes from relinquishing control and trusting God's plan for our lives. Proverbs 3:5-6 (AMP) advises:

"Trust in and rely confidently on the Lord with all your heart, and do not rely on your own insight or understanding. In all your ways know and acknowledge and recognize Him, and He will make your paths straight and smooth [removing obstacles that block your way]."

2. Cultivating a Relationship with the Holy Spirit

To live in the Spirit, we must maintain an intimate relationship with the Holy Spirit. This involves prayer, worship, and studying God's Word. John 14:26 (AMP) promises:

"But the Helper (Comforter, Advocate, Intercessor—Counselor, Strengthener, Standby), the Holy Spirit, whom the Father

will send in My name [in My place, to represent Me and act on My behalf], He will teach you all things. And He will help you remember everything that I have told you."

3. Walking by Faith, Not by Sight
Living in the Spirit requires faith, even when God's purpose is not immediately clear. 2 Corinthians 5:7 (AMP) reminds us:
"For we walk by faith, not by sight [living our lives in a manner consistent with our confident belief in God's promises]."
Trusting in God's promises enables us to move forward confidently, knowing that His plan is perfect.

The Immortal DNA Code: How to Live Eternally in the Spirit

The Immortal DNA Code offers a practical framework for living in alignment with God's eternal purpose:

1. D – Discover Your Divine Purpose
Reflect on your gifts, passions, and life experiences to uncover God's calling. Romans 12:6 (AMP) says:

"Since we have gifts that differ according to the grace given to us, each of us is to use them accordingly."

2. N – Nurture a Spirit-Led Life

Spend time in prayer, worship, and studying God's Word to deepen your relationship with the Holy Spirit. 2 Peter 3:18 (AMP) encourages:

"But grow [spiritually mature] in the grace and knowledge of our Lord and Savior Jesus Christ."

3. A – Align Your Goals with God's Plan

Pursue goals that reflect eternal significance, prioritizing God's kingdom over worldly success. Matthew 6:20 (AMP) instructs:

"But store up for yourselves treasures in heaven, where neither moth nor rust destroys, and where thieves do not break in and steal."

Conclusion

Understanding and living out the truth of Immortal DNA transforms how we view our lives. Our divine purpose, encoded in our spiritual DNA, connects us to eternity and calls us to live in alignment with God's eternal plan. By discovering our purpose, nurturing a Spirit-led life, and aligning our goals with eternal significance, we fulfill the mission God has placed within us.

Philippians 3:14 (AMP) encapsulates the heart of this journey:
"I press on toward the goal to win the [heavenly] prize of the upward call of God in Christ Jesus."

As we embrace the truth of Immortal DNA, we step into a life of eternal significance, living forever as spirits united with God. Let every moment be a reflection of His glory, and let your life echo into eternity.

On Fire for Jesus

The concept of Immortal DNA bridges the physical and spiritual, offering a glimpse into the eternal essence that connects humanity to God's divine purpose. To live forever as an eternal spirit requires embracing the transformative power of living a Christ-centered life. By immersing ourselves in the teachings of Jesus found in the Gospels—Matthew, Mark, Luke, and John—we experience spiritual renewal, favor, and a heart ignited with faith and love. Unlocking the truth of our Immortal DNA means living daily with divine purpose, guided by the Holy Spirit.

The Transformative Power of a Christ-Centered Life

A Christ-centered life begins with an intimate relationship with Jesus, rooted in the study of His life and teachings as recorded in the four Gospels. John 14:6 (AMP) reminds us:

"Jesus said to him, 'I am the [only] Way [to God] and the [real] Truth and the [real] Life; no one comes to the Father but through Me.'"

The Gospels provide the foundation for understanding Jesus' mission, love, and sacrifice. By studying these books, we see how Jesus embodied compassion, humility, and obedience to the Father. This transformative journey starts with:

    1. Reading and Reflecting on the Gospels
The Gospels reveal the heart of Jesus—His love for humanity, His teachings on the kingdom of God, and His ultimate sacrifice on the cross. Matthew 5-7 (the Sermon on the Mount) lays out a blueprint for living a life aligned with God's will.

    2. Emulating Jesus in Daily Life
Jesus calls us not just to know Him but to follow Him actively. Luke 9:23 (AMP) states: *"And He was saying to them all, 'If anyone wishes to follow Me [as My disciple], he must deny himself [set aside selfish interests], and take up his cross daily*

[expressing a willingness to endure whatever may come] and follow Me [believing in Me, conforming to My example in living and, if need be, suffering or perhaps dying because of faith in Me].'"

3. Embracing His Transformative Power
Transformation happens when we align our lives with Christ's example. 2 Corinthians 5:17 (AMP) says:

"Therefore if anyone is in Christ [that is, grafted in, joined to Him by faith in Him as Savior], he is a new creature [reborn and renewed by the Holy Spirit]; the old things [the previous moral and spiritual condition] have passed away. Behold, new things have come [because spiritual awakening brings a new life]."

Experiencing Spiritual Renewal and Favor

Spiritual renewal is a continuous process, made possible through the Holy Spirit. This renewal brings favor and blessings into our lives as we walk in obedience to God's

Word and purpose. Romans 12:2 (AMP) emphasizes:

"And do not be conformed to this world [any longer with its superficial values and customs], but be transformed and progressively changed [as you mature spiritually] by the renewing of your mind [focusing on godly values and ethical attitudes], so that you may prove [for yourselves] what the will of God is, that which is good and acceptable and perfect [in His plan and purpose for you]."

1. Renewing the Mind Through Scripture

Immersing ourselves in the Bible renews our minds, aligning our thoughts with God's truth. Psalm 119:105 (AMP) declares:
"Your word is a lamp to my feet and a light to my path."

2. Walking in God's Favor

Favor is not just material blessings but the presence and guidance of God in our lives. Proverbs 3:3-4 (AMP) explains:
"Do not let mercy and kindness and truth leave you [instead let these qualities define you]; bind them [securely] around your

neck, write them on the tablet of your heart. So find favor and high esteem in the sight of God and man."

3. Spiritual Renewal Through the Holy Spirit

The Holy Spirit works within us to renew and transform our hearts, making us more like Christ. Titus 3:5 (AMP) explains:
"He saved us, not because of any works of righteousness that we have done, but because of His own compassion and mercy, by the cleansing of the new birth (spiritual transformation, regeneration) and renewing by the Holy Spirit."

Manifesting Faith in Daily Actions

True faith is demonstrated through actions rooted in love and compassion. Living a Christ-centered life means manifesting faith in practical ways, reflecting the character of Jesus in all we do. James 2:17 (AMP) reminds us:

"So too, faith, if it does not have works [to back it up], is by itself dead [inoperative and ineffective]."

1. Loving Others as Christ Loved

The hallmark of a Christ-centered life is love. John 13:34-35 (AMP) records Jesus' command:

"I am giving you a new commandment, that you love one another. Just as I have loved you, so you too are to love one another. By this everyone will know that you are My disciples, if you have love and unselfish concern for one another."

2. Serving with Compassion

Jesus modeled servanthood, calling us to follow His example. Matthew 25:40 (AMP) highlights this truth:

"The King will answer and say to them, 'I assure you and most solemnly say to you, to the extent that you did it for one of these brothers of Mine, even the least of them, you did it for Me.'"

3. Sharing the Gospel

Manifesting faith includes sharing the good news of salvation with others. Mark 16:15 (AMP) records Jesus' Great Commission:
"And He said to them, 'Go into all the world and preach the gospel to all creation.'"

Living in the Spirit: Unlocking the Truth About Immortal DNA

Living in the Spirit is the key to activating the Immortal DNA within us. This lifestyle requires yielding to the Holy Spirit, who empowers us to live out God's purpose and fulfill our divine calling. Galatians 5:16 (AMP) instructs:
"But I say, walk habitually in the Holy Spirit [seek Him and be responsive to His guidance], and then you will certainly not carry out the desire of the sinful nature [which responds impulsively without regard for God and His precepts]."

    1. Walking in Daily Obedience
Obedience to God's Word strengthens our spiritual connection and allows us to live in His will. John 15:10 (AMP) says:

"If you keep My commandments and obey My teaching, you will remain in My love, just as I have kept My Father's commandments and remain in His love."

2. Relying on the Holy Spirit

The Holy Spirit provides guidance, comfort, and power to live victoriously. Acts 1:8 (AMP) promises:

"But you will receive power and ability when the Holy Spirit comes upon you; and you will be My witnesses [to tell people about Me] both in Jerusalem and in all Judea, and Samaria, and even to the ends of the earth."

3. Bearing Spiritual Fruit

The evidence of living in the Spirit is the fruit we bear in our lives. Galatians 5:22-23 (AMP) describes:

"But the fruit of the Spirit [the result of His presence within us] is love [unselfish concern for others], joy, [inner] peace, patience [not the ability to wait, but how we act while waiting], kindness, goodness, faithfulness, gentleness, self-control. Against such things there is no law."

The Immortal DNA Code: How to Live Eternally in the Spirit

The Immortal DNA Code provides a practical framework for living in alignment with God's eternal purpose:

1. D – Dedicate Your Life to Christ
Surrender fully to Jesus and commit to living for Him. Romans 10:9 (AMP) assures: *"If you acknowledge and confess with your mouth that Jesus is Lord [recognizing His power, authority, and majesty as God], and believe in your heart that God raised Him from the dead, you will be saved."*

2. N – Nurture a Relationship with God
Spend time in prayer, worship, and studying the Bible to deepen your spiritual connection. Psalm 1:2-3 (AMP) highlights: *"But his delight is in the law of the Lord, and on His law [His precepts and teachings] he [habitually] meditates day and night."*

3. A – Act in Faith and Love
Let your actions reflect your faith and love for others. 1 Corinthians 16:14 (AMP) commands:

"Let everything you do be done in love [motivated and inspired by God's love for us]."

Conclusion

Living on fire for Jesus means embracing the transformative power of a Christ-centered life, experiencing spiritual renewal, and manifesting faith through love and compassion. By immersing ourselves in the teachings of the Gospels and living in the Spirit, we unlock the truth of our Immortal DNA and step into the eternal purpose God has designed for us.

Philippians 3:13-14 (AMP) captures this journey perfectly:
"Brothers and sisters, I do not consider that I have made it my own yet; but one thing I do: forgetting what lies behind and reaching forward to what lies ahead, I press on toward the goal to win the [heavenly] prize of the upward call of God in Christ Jesus."

Let your life reflect His light, your actions demonstrate His love, and your heart burn with a passion to fulfill His eternal plan. This is the essence of living forever as an eternal spirit.

Living in the Spirit

The journey of discovering Immortal DNA begins with a deep understanding of what it means to live in the Spirit. Our physical DNA encodes the instructions for life, but our spiritual DNA connects us to eternal truths and divine purpose. Living in the Spirit is the key to unlocking this immortal connection. It enables us to align with God, reflect His love, and bear fruit that lasts forever. By embracing a Spirit-filled life, we uncover the truth about our immortal nature and learn how to live eternally in the Spirit.

Understanding the Nature of Spiritual Living

Spiritual living transcends the physical realm, calling us to align our hearts and minds with God. It is about surrendering our earthly desires and choosing a life led by the Holy Spirit. Jesus taught that true life is found in the Spirit, as recorded in John 6:63 (AMP):

"It is the Spirit who gives life; the flesh conveys no benefit [it is of no account]. The words I have spoken to you are Spirit and life [providing eternal life]."

1. What is Spiritual Living?

Spiritual living is a life devoted to God, empowered by His Spirit. It means seeking God's will in all things and trusting Him to guide every step. Romans 8:6 (AMP) explains:

"Now the mind of the flesh is death [both now and forever—because it pursues sin]; but the mind of the Spirit is life and peace [the spiritual well-being that comes from walking with God—both now and forever]."

2. The Spirit as a Source of Life

Just as DNA is the blueprint for our physical existence, the Holy Spirit is the blueprint for our spiritual life. He connects us to God's eternal purpose and provides the strength to live according to His will. Ezekiel 36:26-27 (AMP) promises:

"Moreover, I will give you a new heart and put a new spirit within you; and I will remove the heart of stone from your flesh and give you a heart of flesh. I will put My Spirit within you and cause you to walk in My statutes, and you will keep My ordinances and do them."

3. Living Beyond the Temporal

Living in the Spirit means focusing on eternal values rather than temporary pleasures. Colossians 3:2 (AMP) urges: *"Set your mind and keep focused habitually on the things above [the heavenly things], not on things that are on the earth [which have only temporal value]."*

How to Walk in Alignment with the Holy Spirit

Walking in alignment with the Holy Spirit is a continuous, intentional process. It requires daily surrender, active listening, and a commitment to follow God's guidance. Galatians 5:16 (AMP) offers clear instruction:

"But I say, walk habitually in the Holy Spirit [seek Him and be responsive to His guidance], and then you will certainly not carry out the desire of the sinful nature [which responds impulsively without regard for God and His precepts]."

1. Surrender to God's Will

The first step in walking with the Spirit is surrendering control of your life to God. Proverbs 3:5-6 (AMP) reminds us:

"Trust in and rely confidently on the Lord with all your heart and do not rely on your own insight or understanding. In all your ways know and acknowledge and recognize Him, and He will make your paths straight and smooth [removing obstacles that block your way]."

2. Listen to the Spirit's Voice

The Holy Spirit speaks to us through God's Word, prayer, and inner conviction. Isaiah 30:21 (AMP) assures:

"Your ears will hear a word behind you, saying, 'This is the way, walk in it,' whenever you turn to the right or to the left."

3. Obey Promptings Without Hesitation

Living in alignment with the Spirit requires immediate obedience. Delayed obedience often leads to missed opportunities to fulfill God's will. Acts 8:29-30 (AMP) illustrates this principle in the story of Philip and the Ethiopian eunuch:

"Then the [Holy] Spirit said to Philip, 'Go up and join this chariot.' Philip ran up and heard the man reading the prophet Isaiah, and asked, 'Do you understand what you are reading?'"

4. Cultivate Spiritual Disciplines

Prayer, fasting, worship, and Bible study are essential practices for maintaining alignment with the Spirit. Matthew 6:6 (AMP) encourages:

"But when you pray, go into your most private room, close the door and pray to your Father who is in secret; and your Father who sees [what is done] in secret will reward you."

The Fruits of Living a Spirit-Filled Life

When we live in the Spirit, our lives produce visible evidence of God's presence and work within us. These fruits reflect the character of Jesus and bring glory to God.
Galatians 5:22-23 (AMP) describes the fruits of the Spirit:
"But the fruit of the Spirit [the result of His presence within us] is love [unselfish concern for others], joy, [inner] peace, patience [not the ability to wait, but how we act while waiting], kindness, goodness, faithfulness, gentleness, self-control. Against such things there is no law."

    1. Love

Living in the Spirit allows us to love unconditionally, just as God loves us. 1 John 4:7 (AMP) states:

"Beloved, let us [unselfishly] love and seek the best for one another, for love is from God; and everyone who loves [others] is born of God and knows God [through personal experience]."

2. Joy

Joy is a deep, abiding sense of happiness that comes from knowing God and trusting His plan. Psalm 16:11 (AMP) proclaims:

"You will show me the path of life; in Your presence is fullness of joy; in Your right hand there are pleasures forevermore."

3. Peace

The peace of the Spirit surpasses all understanding, calming our hearts even in the midst of chaos. Philippians 4:7 (AMP) affirms:

"And the peace of God [that peace which reassures the heart, that peace] which

transcends all understanding, [that peace which] stands guard over your hearts and your minds in Christ Jesus [is yours]."

4. Patience, Kindness, and Goodness

These fruits enable us to reflect God's character in our interactions with others, showing grace and compassion. Ephesians 4:32 (AMP) instructs:
"Be kind and helpful to one another, tender-hearted [compassionate, understanding], forgiving one another [readily and freely], just as God in Christ also forgave you."

5. Faithfulness, Gentleness, and Self-Control

These traits help us maintain a steadfast commitment to God, respond with humility, and resist the temptations of the flesh. 2 Timothy 1:7 (AMP) reminds us:
"For God did not give us a spirit of timidity or cowardice or fear, but [He has given us a spirit] of power and of love and of sound judgment and personal discipline [abilities

that result in a calm, well-balanced mind and self-control]."

Living in Spirit: Unlocking the Truth About Immortal DNA

Living in the Spirit is the key to unlocking the truth about our Immortal DNA. When we surrender to the Spirit, align with His guidance, and bear His fruit, we begin to live eternally in the Spirit. Romans 8:11 (AMP) assures us of this eternal connection: *"And if the Spirit of Him who raised Jesus from the dead lives in you, He who raised Christ Jesus from the dead will also give life to your mortal bodies through His Spirit who lives in you."*

The Immortal DNA Code: How to Live Eternally in the Spirit

The Immortal DNA Code serves as a practical guide to living a Spirit-filled life and walking in God's eternal purpose:

1. D – Dedicate Your Life to God

Fully commit to God's will and purpose for your life. Matthew 22:37 (AMP) commands:

"And Jesus replied to him, 'You shall love the Lord your God with all your heart, and with all your soul, and with all your mind.'"

2. N – Nurture Your Spiritual Connection

Build an intimate relationship with God through prayer, worship, and scripture. Psalm 119:105 (AMP) declares:

"Your word is a lamp to my feet and a light to my path."

3. A – Align with the Holy Spirit

Actively seek the Spirit's guidance in every decision and action. Galatians 5:25 (AMP) urges:

"If we claim to live by the Holy Spirit, we must also walk by the Spirit [with personal integrity, godly character, and moral courage—our conduct empowered by the Holy Spirit]."

By following this code, we align our lives with God's eternal plan and begin to live as His eternal children.

Conclusion

Living in the Spirit is the essence of unlocking the truth about Immortal DNA. It is a journey of surrender, alignment, and transformation, made possible through the guidance of the Holy Spirit. By walking in the Spirit, bearing His fruit, and embracing our divine purpose, we step into the reality of living eternally in God's presence.

2 Corinthians 4:18 (AMP) perfectly encapsulates this eternal perspective:
"So we look not at the things which are seen, but at the things which are unseen; for the things which are visible are temporal [just brief and fleeting], but the things which are invisible are everlasting and imperishable."

Let your life be a testament to the eternal power of the Spirit within you. Unlock your Immortal DNA and live forever as an eternal spirit in alignment with God's divine purpose.

Your Pending Creations

The concept of Immortal DNA is more than just a metaphor for eternity; it is a divine truth embedded in the very essence of who we are. Created in the image of God, we are designed to be co-creators with Him, shaping our world, leaving a legacy, and aligning our efforts with His divine plan. The topic of "Your Pending Creations" invites us to explore the immense untapped potential within us, recognize our identity as divine co-creators, and understand how we can manifest spiritual intentions into reality.

Through the lens of the Bible, this journey unveils the profound connection between our spiritual DNA and the eternal purpose God has placed within each of us. Living forever as an eternal spirit means actively engaging with this purpose, harnessing our creative power, and ensuring our creations align with God's Word for divine favor and everlasting impact.

The Untapped Potential Within You to Create and Influence

God has instilled in every person a wellspring of creative potential, a reflection of His own divine creativity. Genesis 1:27 (AMP) declares:
"So God created man in His own image, in the image and likeness of God He created him; male and female He created them."
This foundational truth reminds us that we are not only recipients of God's creativity but also carriers of His ability to create.

   1. The Power of Potential
Your spiritual DNA contains the seeds of untapped potential. Whether it's a business, a book, a song, or a relationship, God has placed within you unique gifts that are meant to glorify Him and serve others. 2 Timothy 1:6 (AMP) exhorts:
"That is why I remind you to fan into flame the gracious gift of God [that inner fire—the special endowment] which is in you."

   2. The Ripple Effect of Influence

When we use our gifts, we create ripples that extend far beyond what we can see. A single word of encouragement, a heartfelt song, or a transformative invention can change the trajectory of lives. Jesus highlighted this potential in Matthew 5:16 (AMP):

"Let your light shine before men in such a way that they may see your good deeds and moral excellence, and [recognize and honor and] glorify your Father who is in heaven."

3. Cultivating Your Potential

Like any seed, your potential must be nurtured. This involves seeking God's guidance, stepping out in faith, and diligently developing your skills. Colossians 3:23 (AMP) reminds us:

"Whatever you do [whatever your task may be], work from the soul [that is, put in your very best effort], as [something done] for the Lord and not for men."

Recognizing the Divine Co-Creator Within

As bearers of God's image, we are called to be co-creators with Him. This divine partnership is not limited to physical creations like children but extends to everything we can dream and manifest, from businesses and books to songs, art, and acts of service.

1. What Can You Create?

The possibilities are limitless when we recognize our role as co-creators with God. Here are some examples:

- A Child: Parenting is one of the most profound acts of co-creation. Psalm 127:3 (AMP) celebrates:

"Behold, children are a heritage and gift from the Lord, the fruit of the womb a reward."

- A Business: Building a business aligned with God's principles can bless countless lives and reflect His glory. Proverbs 16:3 (AMP) advises:

"Commit your works to the Lord [submit and trust them to Him], and your plans will succeed [if you respond to His will and guidance]."

- A Work of Art: Artistic creations can inspire, heal, and communicate God's truth. Exodus 31:3 (AMP) reveals:

"I have filled him with the Spirit of God in wisdom and skill, in understanding and intelligence, in knowledge, and in all kinds of craftsmanship."

2. Aligning Your Creations with God's Word

To unlock the full favor of God, it is essential to align your creations with His Word. This means seeking His guidance, dedicating your efforts to His glory, and ensuring your motives are pure. Psalm 37:5-6 (AMP) encourages:

"Commit your way to the Lord; trust in Him also and He will do it. He will make your righteousness [your pursuit of right standing with God] like the light, and your judgment like the [shining of the] noonday sun."

3. Receiving Divine Favor

When your creations align with God's purposes, they carry His blessing and favor. Proverbs 3:9-10 (AMP) promises:

"Honor the Lord with your wealth and with the first fruits of all your crops (income); then your barns will be abundantly filled and your vats will overflow with new wine."

Manifesting Spiritual Intentions Into Reality

Manifestation, in the biblical sense, is not about selfish desires but about bringing God's will into fruition through faith and action.

1. Aligning Intentions with God's Will
The first step in manifesting spiritual intentions is ensuring they align with God's will. James 4:3 (AMP) warns:
"You ask [God for something] and do not receive it, because you ask with wrong motives [out of selfishness or with an unrighteous agenda], so that [when you get what you want] you may spend it on your [hedonistic] desires."

2. Faith as the Foundation
Faith is the catalyst for manifestation. Hebrews 11:1 (AMP) defines faith:

"Now faith is the assurance (title deed, confirmation) of things hoped for (divinely guaranteed), and the evidence of things not seen [the conviction of their reality—faith comprehends as fact what cannot be experienced by the physical senses]."

3. Taking Inspired Action

Faith without action is incomplete. James 2:26 (AMP) declares:

"For just as the body without the spirit is dead, so also faith without works is dead."

Inspired action involves listening to God's guidance and taking steps toward fulfilling His purpose for your creations.

4. Trusting God's Timing

Manifestation requires patience and trust in God's perfect timing. Ecclesiastes 3:11 (AMP) assures us:

"He has made everything beautiful and appropriate in its time."

Living in Spirit: Unlocking the Truth About Immortal DNA

Living in the Spirit is the key to understanding your Immortal DNA and unlocking your divine potential. It is a life of alignment with God's will, a commitment to His purposes, and a willingness to co-create with Him. Romans 8:14 (AMP) declares: *"For all who are allowing themselves to be led by the Spirit of God are sons of God."*

The Immortal DNA Code: How to Live Eternally in the Spirit

The Immortal DNA Code serves as a guide for unlocking your divine potential and living in alignment with God's eternal purpose:

    1. D – Dedicate Your Creations to God
Submit every creative endeavor to God's guidance and glory. Proverbs 16:9 (AMP): *"A man's mind plans his way [as he journeys through life], but the Lord directs his steps and establishes them."*
    2. N – Nurture Your God-Given Gifts
Cultivate your talents and develop your skills to honor God. Matthew 25:21 (AMP):

"His master said to him, 'Well done, good and faithful servant. You have been faithful and trustworthy over a little; I will put you in charge of many things; share in the joy of your master.'"

3. A – Align Your Intentions with God's Will

Seek God's purpose in all you create, ensuring your efforts reflect His love and truth. Jeremiah 29:11 (AMP):

"For I know the plans and thoughts that I have for you,' says the Lord, 'plans for peace and well-being and not for disaster, to give you a future and a hope."

By following this code, you unlock your potential to create and influence, leaving a legacy that glorifies God and echoes through eternity.

Conclusion

Your pending creations are not just earthly endeavors but divine opportunities to glorify God and fulfill His purpose. By recognizing the co-creator within you, aligning your creations with His Word, and manifesting your spiritual intentions through faith and action, you step into your true identity as a child of God with Immortal DNA.

Ephesians 2:10 (AMP) sums up this divine calling:
"For we are His workmanship [His own master work, a work of art], created in Christ Jesus [reborn from above—spiritually transformed, renewed, ready to be used] for good works, which God prepared [for us] beforehand [taking paths which He set], so that we would walk in them [living the good life which He prearranged and made ready for us]."

Live forever as an eternal spirit by embracing your Immortal DNA and co-creating with God to manifest His glory on earth.

Making Ripples in Time

The essence of Immortal DNA lies in its ability to transcend time, carrying the imprint of divine creativity and purpose through generations. Within each of us is a spark of eternity, a spiritual DNA encoded with the potential to influence not only our lifetime but the lives of those who come after us. The concept of Making Ripples in Time illuminates how small, intentional actions can create a lasting impact that echoes across generations. By living in alignment with God's will and understanding the eternal significance of our choices, we unlock the truth about Immortal DNA and live forever as eternal spirits.

How Small Actions Today Echo into Eternity

Every action we take has consequences that ripple through time, akin to the butterfly effect, a term used to describe how small, seemingly insignificant actions can lead to monumental outcomes. This phenomenon mirrors the spiritual realm, where even the smallest act of kindness or faithfulness can reverberate throughout eternity.

Galatians 6:9 (AMP) encourages:
"Let us not grow weary or become discouraged in doing good, for at the proper time we will reap, if we do not give in."

1. The Power of Small Actions

A kind word, a single prayer, or a moment of generosity may appear insignificant, but in the spiritual realm, these acts carry eternal weight. Jesus emphasized this in Matthew 25:40 (AMP):
"The King will answer and say to them, 'I assure you and most solemnly say to you, to the extent that you did it for one of these brothers of Mine, even the least of them, you did it for Me.'"

2. The Butterfly Effect of Faith

Just as the flap of a butterfly's wings can influence distant weather systems, our small steps of obedience and love can trigger spiritual transformations far beyond our comprehension. Consider the widow's offering in Mark 12:43-44 (AMP):

"Truly I say to you, this poor widow put in more than all the contributors to the treasury; for they all contributed from their surplus, but she from her poverty put in all she had, all she had to live on."

The Concept of Spiritual Legacy

A spiritual legacy is not measured by wealth or material success but by the eternal impact of a life lived in alignment with God's purpose. To leave a spiritual legacy is to create something that endures through generations, inspiring faith, love, and purpose in those who follow.

1. What is a Spiritual Legacy?

A spiritual legacy is the eternal imprint left by our words, actions, and creations. It reflects how we lived, who we served, and how we glorified God. Proverbs 13:22 (AMP) affirms:

"A good man leaves an inheritance to his children's children, and the wealth of the sinner is stored up for [the hands of] the righteous."

2. Lasting Creations Across Generations

- Faith-Based Teachings: Sharing God's Word with others creates a foundation of faith that can guide generations.

- Acts of Love: Compassionate deeds inspire others to live in the Spirit and continue spreading love.

- Tangible Creations: Writing a book, building a ministry, or founding a charity are tangible ways to leave a spiritual legacy that endures.

3. Biblical Examples of Spiritual Legacy

Abraham is a powerful example of a spiritual legacy. His faith and obedience established a covenant that blessed countless generations. Genesis 17:7 (AMP) states:

"I will establish My covenant between Me and you and your descendants after you throughout their generations for an everlasting covenant, to be God to you and to your descendants after you."

Creating Generational Ripple Effects

The ripple effect of a life lived with love and purpose transcends time, influencing not only immediate descendants but also communities and nations.

1. The Ripple Effect of Love

Love is the most powerful force for creating ripples in time. When we love others selflessly, we reflect God's eternal love, which transforms lives. 1 Corinthians 13:13 (AMP) declares:

"And now there remain faith [abiding trust in God and His promises], hope [confident expectation of eternal salvation], love

[unselfish love for others growing out of God's love for me]. These three [the choicest graces]; but the greatest of these is love."

2. Purposeful Living for Generational Impact

Aligning your life with God's purpose ensures that your actions are divinely guided to create lasting change. Jeremiah 29:11 (AMP) reminds us:

"For I know the plans and thoughts that I have for you,' says the Lord, 'plans for peace and well-being and not for disaster, to give you a future and a hope."

3. Practical Steps to Create Ripples

• Mentorship: Invest in the spiritual growth of others by sharing your faith journey.

• Service: Actively contribute to your community through acts of kindness and service.

• Prayer: Pray for future generations, knowing that God hears and answers beyond your lifetime.

Living in Spirit: Unlocking the Truth About Immortal DNA

Living in the Spirit means understanding and embracing your divine calling to influence the world with love and purpose. By aligning your actions with God's will, you live as an eternal spirit, leaving ripples that carry His glory into eternity.

1. Walking in the Spirit

Walking in the Spirit requires daily surrender to God's guidance. Galatians 5:16 (AMP) teaches:
"But I say, walk habitually in the [Holy] Spirit [seek Him and be responsive to His guidance], and then you will certainly not carry out the desire of the sinful nature."

2. Bearing the Fruits of the Spirit

The evidence of living in the Spirit is seen in the fruit you bear. Galatians 5:22-23 (AMP) describes these fruits:

"But the fruit of the Spirit [the result of His presence within us] is love [unselfish concern for others], joy, [inner] peace, patience [not the ability to wait, but how we act while waiting], kindness, goodness, faithfulness, gentleness, self-control."

3. Recognizing Your Eternal Identity

Living in the Spirit reveals your true identity as an eternal being.
Romans 8:16-17 (AMP) declares:
"The Spirit Himself testifies and confirms together with our spirit [assuring us] that we [believers] are children of God. And if [we are His] children, [then we are His] heirs also: heirs of God and fellow heirs with Christ [sharing His spiritual blessing and inheritance]."

The Immortal DNA Code: How to Live Eternally in the Spirit

The Immortal DNA Code is a guide for living a life that creates ripples of eternal significance:

1. D – Dedicate Every Action to God

Live intentionally, committing your words and deeds to God's glory. Colossians 3:17 (AMP):

"Whatever you do [no matter what it is] in word or deed, do everything in the name of the Lord Jesus [and in dependence on Him], giving thanks to God the Father through Him."

2. N – Nurture Your Spiritual Legacy

Invest in relationships, faith-based teachings, and creations that align with God's eternal purpose. Psalm 78:4 (AMP):

"We will not hide them from their children, but [we will] tell to the generation to come the praiseworthy deeds of the Lord, and [tell of] His great might and power and the wonderful works that He has done."

3. A – Act with Eternal Purpose

Recognize the significance of every choice you make, knowing it carries

eternal consequences. Matthew 6:20 (AMP):
"But store up for yourselves treasures in heaven, where neither moth nor rust destroys, and where thieves do not break in and steal."

Conclusion

Understanding Immortal DNA means embracing your divine potential to create ripples of love, faith, and purpose that resonate through eternity. By living in the Spirit, dedicating your actions to God, and leaving a spiritual legacy, you unlock the truth of your eternal identity and influence the world for generations to come.

Ephesians 2:10 (AMP) encapsulates this divine calling:
"For we are His workmanship [His own master work, a work of art], created in Christ Jesus [reborn from above—spiritually transformed, renewed, ready to be used] for good works, which God prepared [for us]

beforehand [taking paths which He set], so that we would walk in them [living the good life which He prearranged and made ready for us]."

Live forever as an eternal spirit by understanding your Immortal DNA and using your life to make ripples that glorify God and inspire future generations.

Becoming an Eternal Star

The concept of Immortal DNA reveals the eternal nature of our souls, our unique spiritual identity, and the divine purpose encoded within us. By living in the Spirit and embracing the transformative power of Jesus Christ, we unlock the truth about our eternal destiny. The idea of Becoming an Eternal Star offers a profound metaphor for the spiritual journey—illuminating others with the divine light of Christ, fulfilling our God-given purpose, and living forever as an eternal spirit.

Embracing Your Unique Spiritual Identity

Each individual is a masterpiece, crafted by God with intention and purpose. Your spiritual DNA is uniquely designed to reflect God's glory. Psalm 139:14 (AMP) reminds us:
"I will give thanks and praise to You, for I am fearfully and wonderfully made;

Wonderful are Your works, and my soul knows it very well."

1. The Heart Container Filled with Jesus
The metaphor of the heart as a container reflects its capacity to be filled with love, faith, and divine purpose. When your heart is overflowing with Jesus, your life transforms into a reflection of Heaven on Earth. This divine fulfillment is beautifully captured in 2 Corinthians 4:6 (AMP):
"For God, who said, 'Let light shine out of darkness,' is the One who has shone in our hearts to give us the Light of the knowledge of the glory and majesty of God [clearly revealed] in the face of Christ."

2. The Eternal Star Hung in the Sky
In this metaphor, the moment your heart is filled with Christ, an eternal star—symbolizing your spiritual legacy—is hung in the heavens. This star represents your unique contribution to God's Kingdom and serves as a beacon of hope and faith for others. Daniel 12:3 (AMP) speaks of this eternal illumination:

"Those who are wise [believers] will shine like the brightness of the expanse of heaven, and those who lead many to righteousness [will shine] like the stars forever and ever."

Faith and Purpose: Guiding You to Eternal Significance

Faith is the foundation upon which your spiritual identity is built, and purpose is the divine mission that directs your life. Together, they guide you toward eternal significance.

    1. The Visible Presence of the Holy Spirit
When faith and purpose align, the Holy Spirit becomes a visible and active presence in your life. Galatians 5:25 (AMP) encourages:
"If we claim to live by the [Holy] Spirit, we must also walk by the Spirit [with personal integrity, godly character, and moral courage—our conduct empowered by the Holy Spirit]."
    2. Living with Eternal Significance

Eternal significance is achieved when your life reflects God's eternal glory. This requires a surrender of earthly ambitions and a commitment to divine purpose. Colossians 3:2 (AMP) reminds us:
"Set your mind and keep focused habitually on the things above [the heavenly things], not on things that are on the earth [which have only temporal value]."

3. Transformative Power of Purpose
Purpose aligns your actions with God's will, enabling you to fulfill your role in His Kingdom. Jeremiah 1:5 (AMP) declares:
"Before I formed you in the womb I knew you [and approved of you as My chosen instrument], and before you were born I consecrated you [to Myself as My own]; I have appointed you as a prophet to the nations."

Illuminating Others Through Your Divine Light

The light of Christ within you is not meant to be hidden but to shine brightly,

illuminating the path for others.
Matthew 5:14-16 (AMP) declares:
"You are the light of [Christ to] the world. A city set on a hill cannot be hidden; nor does anyone light a lamp and put it under a basket, but on a lampstand, and it gives light to all who are in the house. Let your light shine before men in such a way that they may see your good deeds and moral excellence, and [recognize and honor and] glorify your Father who is in heaven."

1. Radiating the Love of Christ

When your heart is filled with Jesus, your life becomes a radiant reflection of His love. This light has the power to draw others closer to God, inspiring faith and transformation. 1 John 4:16 (AMP) emphasizes:
"We have come to know [by personal observation and experience], and have believed [with deep, consistent faith] the love which God has for us. God is love, and the one who abides in love abides in God, and God abides continually in him."

2. Sharing the Light Through Actions

Every action taken in love and obedience to God creates ripples of hope and faith. Acts of kindness, service, and prayer become vessels through which Christ's light is shared. Ephesians 2:10 (AMP) reminds us: *"For we are His workmanship [His own master work, a work of art], created in Christ Jesus [reborn from above—spiritually transformed, renewed, ready to be used] for good works, which God prepared [for us] beforehand [taking paths which He set], so that we would walk in them [living the good life which He prearranged and made ready for us]."*

3. Inspiring Others to Shine
Your light not only illuminates the way but also inspires others to discover and embrace their own divine light. This communal illumination builds the Kingdom of God on Earth, fulfilling His purpose for humanity.

Living in Spirit: Unlocking the Truth About Immortal DNA

Living in the Spirit involves a daily commitment to walking in faith, surrendering to God's guidance, and allowing the Holy Spirit to transform your life. Romans 8:14 (AMP) explains:
"For all who are allowing themselves to be led by the Spirit of God are sons of God."

1. The Process of Transformation
Transformation begins with the renewal of your mind and heart through the Word of God. Romans 12:2 (AMP) encourages:
"And do not be conformed to this world [any longer with its superficial values and customs], but be transformed and progressively changed [as you mature spiritually] by the renewing of your mind [focusing on godly values and ethical attitudes], so that you may prove [for yourselves] what the will of God is, that which is good and acceptable and perfect [in His plan and purpose for you]."

2. The Fruits of Living in the Spirit
A Spirit-filled life is evidenced by the fruit it bears, including love, joy, peace, and

faithfulness. Galatians 5:22-23 (AMP) describes:

"But the fruit of the Spirit [the result of His presence within us] is love [unselfish concern for others], joy, [inner] peace, patience [not the ability to wait, but how we act while waiting], kindness, goodness, faithfulness, gentleness, self-control."

   3. The Eternal Perspective

Living in the Spirit grants you an eternal perspective, where earthly struggles are seen as opportunities for spiritual growth and divine purpose. 2 Corinthians 4:17-18 (AMP) assures us:

"For our momentary, light distress [this passing trouble] is producing for us an eternal weight of glory [a fullness] beyond all measure [surpassing all comparisons, a transcendent splendor and an endless blessedness]! So we look not at the things which are seen, but at the things which are unseen; for the things which are visible are temporal [just brief and fleeting], but the things which are invisible are everlasting and imperishable."

The Immortal DNA Code: How to Live Eternally in the Spirit

The Immortal DNA Code provides a framework for living as an eternal spirit, reflecting God's glory and fulfilling your divine purpose:

1. D – Discover Your Spiritual Identity

Embrace your uniqueness as God's creation, designed for eternal significance. Ephesians 1:4 (AMP):

"Just as [in His love] He chose us in Christ [actually selected us for Himself as His own] before the foundation of the world, so that we would be holy [that is, consecrated, set apart for Him, purpose-driven] and blameless in His sight."

2. N – Nurture Your Relationship with Christ

Allow Jesus to fill your heart completely, transforming your life and guiding your purpose. John 15:5 (AMP):

"I am the Vine; you are the branches. The one who remains in Me and I in him bears

much fruit, for [otherwise] apart from Me [that is, cut off from vital union with Me] you can do nothing."

3. A – Align Your Actions with God's Will
Shine your light by living a life of love, faith, and service, creating ripples of eternal significance. Matthew 5:16 (AMP):
"Let your light shine before men in such a way that they may see your good deeds and moral excellence, and [recognize and honor and] glorify your Father who is in heaven."

Conclusion

Becoming an Eternal Star means embracing your spiritual identity, living with purpose, and illuminating others with the divine light of Christ. By allowing Jesus to fill your heart, walking in alignment with the Holy Spirit, and creating ripples of love and faith, you unlock the truth of Immortal DNA and live forever as an eternal spirit.

Philippians 2:15 (AMP) captures the essence of this transformation:

"So that you may prove yourselves to be blameless and guileless, innocent and uncontaminated, children of God without blemish in the midst of a morally crooked and spiritually perverted generation, among whom you are seen as bright lights [beacons shining out clearly] in the world."

Live as an eternal star, reflecting God's glory and inspiring others to join His Kingdom. This is the fulfillment of your Immortal DNA and the path to eternal life in the Spirit.

Legacy of Immortal DNA

The concept of Immortal DNA speaks to the spiritual truths encoded within us, extending beyond physical biology into eternal purpose and destiny. This legacy is passed on not through mere genetics, but through the choices, actions, and spiritual imprints we leave behind for generations. By embracing the immortality of the spirit through faith, purpose, and creative expressions, we ensure our impact endures long after our earthly journey. The Legacy of Immortal DNA is the story of how we live eternally by living in the Spirit and influencing the world for God's Kingdom.

Passing on Spiritual Wisdom and Purpose

The core of the Immortal DNA Legacy is the transmission of spiritual wisdom. Creative expressions, such as books, music, art, films, or even spoken words, are eternal imprints of our divine purpose. These creations serve as vessels for God's truth, carrying His love and wisdom across time.

Proverbs 22:6 (AMP) emphasizes the power of nurturing others spiritually:
"Train up a child in the way he should go [teaching him to seek God's wisdom and will for his abilities and talents]; even when he is old he will not depart from it."

    1. Creative Media as Eternal Imprints
Creative media offers a way to encode spiritual truths in forms that resonate across cultures and generations. For example, a song infused with God's love can bring healing to someone decades after its creation. By aligning our creations with God's Word, we ensure they carry divine favor and eternal significance.

    2. Living as a Teacher of Purpose
Each of us is a teacher, whether through words, actions, or creative works. Sharing your testimony, faith, and the lessons God has taught you can inspire countless lives. Matthew 28:19-20 (AMP) reminds us of this calling:

"Go therefore and make disciples of all the nations [help the people to learn of Me, believe in Me, and obey My words], baptizing them in the name of the Father and of the Son and of the Holy Spirit, teaching them to observe everything that I have commanded you."

   3. Eternal Imprints Through Love and Obedience

Spiritual wisdom is most powerfully passed on through love and obedience to God. The legacy you leave through your relationships, service, and faithfulness reflects the heart of Christ, leaving a mark on every soul you touch.

Eternal Choices and Their Impact on Generations

Every choice you make has the potential to influence eternity, not only for yourself but for those who come after you. This ripple effect of spiritual decisions creates a legacy that spans generations.

   1. Choosing Eternity Over Temporality

The decisions rooted in faith, love, and obedience to God have eternal consequences. Colossians 3:23-24 (AMP) encourages:

"Whatever you do [whatever your task may be], work from the soul [that is, put in your very best effort], as [something done] for the Lord and not for men, knowing [with all certainty] that it is from the Lord [not from men] that you will receive the inheritance which is your [greatest] reward. It is the Lord Christ whom you [actually] serve."

2. Impacting Generations Through Example

Your life serves as a testimony to others. Whether raising a family, leading a community, or creating meaningful work, your faithfulness to God inspires others to follow Him. Deuteronomy 7:9 (AMP) highlights the generational blessings of obedience:

"Know therefore that the Lord your God, He is God, the faithful God, who is keeping His covenant and His steadfast lovingkindness

to a thousand generations with those who love Him and keep His commandments."

   3. Creating Ripple Effects Through Spiritual Acts

A single act of kindness, faith, or creativity can create ripples that echo through eternity. Whether you inspire someone to pursue their God-given calling or share the gospel, your influence extends far beyond what you can see.

Recognizing the Immortality of Spirit Through Faith in Action

Faith is the bridge that connects our earthly lives to the eternal. It is through faith that we recognize the immortality of the spirit and align our actions with God's divine purpose.

   1. Faith as the Foundation of Immortality
Hebrews 11:1 (AMP) defines faith as:
"Now faith is the assurance [title deed, confirmation] of things hoped for (divinely guaranteed), and the evidence of things not seen [the conviction of their reality—faith

comprehends as fact what cannot be experienced by the physical senses]."
Faith assures us that our actions, grounded in God's will, carry eternal weight.

2. Walking in Faithful Action
Living in faith means stepping out in obedience to God, even when the outcome is uncertain. James 2:17 (AMP) reminds us: *"So too, faith, if it does not have works [to back it up], is by itself dead [inoperative and ineffective]."*
Faith is made alive through action, and these actions leave a lasting imprint on the world.

3. Living as a Co-Creator with God
Recognizing the immortality of your spirit empowers you to live as a co-creator with God. By aligning your actions, choices, and creations with His will, you participate in His eternal plan.

Living in Spirit: Unlocking the Truth About Immortal DNA

Living in the Spirit is the key to unlocking the truth about your Immortal DNA. It is the daily practice of aligning with God's presence, walking in obedience, and manifesting His Kingdom on Earth.

1. Walking in Alignment with the Holy Spirit

The Holy Spirit empowers and guides us in living a life of eternal significance. Romans 8:14 (AMP) explains:

"For all who are allowing themselves to be led by the Spirit of God are sons of God."

This alignment with the Holy Spirit allows us to live in harmony with God's will.

2. Bearing the Fruits of the Spirit

A Spirit-filled life is marked by the fruits it bears, as described in Galatians 5:22-23 (AMP):

"But the fruit of the Spirit [the result of His presence within us] is love [unselfish concern for others], joy, [inner] peace, patience [not the ability to wait, but how we act while waiting], kindness, goodness, faithfulness, gentleness, self-control."

These fruits are evidence of a life lived in communion with God and leave a legacy of love and transformation.

3. Manifesting God's Kingdom on Earth
Living in the Spirit involves bringing God's Kingdom into the here and now through faith, service, and creativity. Matthew 6:10 (AMP) calls us to pray:
"Your kingdom come, Your will be done on earth as it is in heaven."

The Immortal DNA Code: How to Live Eternally in the Spirit

The Immortal DNA Code offers a practical guide for living eternally in the Spirit, ensuring your life and legacy reflect God's eternal glory:

1. D – Discern Your Purpose
Seek God's guidance to uncover your divine purpose. Proverbs 19:21 (AMP) advises:
"Many plans are in a man's mind, but it is the Lord's purpose for him that will stand [be carried out]."

2. N – Nurture Your Spirit

Cultivate a deep relationship with God through prayer, worship, and the Word. John 15:7 (AMP) promises:

"If you remain in Me and My words remain in you [that is, if we are vitally united and My message lives in your heart], ask whatever you wish and it will be done for you."

3. A – Act with Eternal Intent

Live with the awareness that your actions carry eternal significance. Matthew 6:33 (AMP) encourages:

"But first and most importantly seek (aim at, strive after) His kingdom and His righteousness [His way of doing and being right—the attitude and character of God], and all these things will be given to you also."

Conclusion

The Legacy of Immortal DNA is a testimony to the eternal impact of a life lived in faith, purpose, and love. By passing on spiritual wisdom, making choices with eternal significance, and recognizing the immortality of the spirit, we ensure our lives echo through generations. Living in the Spirit unlocks the truth of Immortal DNA, empowering us to co-create with God and manifest His Kingdom on Earth.

As 2 Timothy 4:7-8 (AMP) declares:
"I have fought the good and worthy and noble fight, I have finished the race, I have kept the faith [firmly guarding the gospel against error]. In the future there is reserved for me the victor's crown of righteousness [for being right with God and doing right], which the Lord, the righteous Judge, will award to me on that great day—and not to me only, but also to all those who have loved and longed for and welcomed His appearing."

Live with the confidence that your Immortal DNA connects you to God's eternal plan, and leave a legacy that shines as a testament to His glory for generations to come.

About Scott Perdue

Scott Perdue is a dynamic entrepreneur, author, and community leader with a life rooted in faith, family, and service. A devoted Christian, Scott has been married for over 20 years and is the proud father of four children—two girls and two boys. His passion for personal development and spiritual growth is reflected in his prolific writing career, having authored over 100 books, most of which focus on self-help and Christian themes. His books have touched the lives of countless readers seeking guidance on how to lead a fulfilling, faith-centered life.

For over 15 years, Scott has been a dedicated member of GUTS Church, a place he fondly refers to as "It Takes GUTS to Serve the Lord." His service to the church and community extends beyond attendance; he spent six years as a representative for the GUTS Food Bank, where he managed the movement of wholesale goods to help those in need. Scott also led a successful Maximized Manhood study group based on Edwin Cole's teachings, further exemplifying his commitment to fostering spiritual growth among men.

An accomplished entrepreneur, Scott has started and operated over 30 businesses, ranging from pest control to contracting. He is the founder of Universal Bug Man, a pest control service where Scott earned a reputation as a "pest control superhero." His entrepreneurial ventures include Tulsa Furniture Wholesale, Tulsa Auction Spot, and Builderhaus Unlimited, among others. Scott's business acumen extends to the health and wellness industry, where his company HCG Medical helped over 20,000 clients lose weight, generating over $6.5 million in sales in its best year.

Scott Perdue is a man of many talents, driven by his faith and dedication to serving others through his varied enterprises and writing.

Follow Scott Perdue on YouTube, Facebook & Visit UniversalWholesaleStore.com

Published Books by Scott Perdue (Buy Today on Amazon) ⬇

Christian Books by Scott Perdue:

Biblical Entrepreneur Leadership: Amplified Leverage Business Skills Book & Workbook

Biblical Men's Leadership Skills: Becoming an Amplified Christian Superstar Book & Workbook

Unleashing Biblical Manhood: Taking Ground Like a Warrior Book & Workbook

Promised Land Leadership: Leading an Army Like Joshua

Wilderness Wisdom of Moses: Timeless Life-Changing Leadership Lessons

Rules of Christianity According to Paul Book & Workbook

Provisional Miracles of Jesus: Provision through Supernatural Means Book & Workbook

Kingdom Money: Unlocking Biblical Secrets to Financial Success

The King's Highway: Lean into Jesus for Accelerated Success

Walk in the Works of the Lord: An Amplified Passion Understanding

God's River: Getting into the Kingdom Family Flow

Forgiven & Unoffendable: The Power of Walking Righteously

God is Real: Knowing the Spirit - A Journey Through Faith, Miracles, and Divine Presence

The Gift of Light: A Journey of Spiritual Growth for Life Expansion

On Fire For Jesus: Bring Plasma Energy to Your Heart Pump

Immortal DNA: Living Forever as an Eternal Spirit

Earth is God's Beach Ball: Celebrate the Legacy of Joyful Living

Faith in the Wilderness: Biblical Lessons for Strength and Spiritual Growth

Living on Purpose: A Comprehensive Guide to a Meaningful and Fulfilling Life

Praying for Others: Unlocking your God-Given Authority to Change Lives

Speaking in Tongues: Snippets of Life Improvement Code

Be Fruitful and Multiply: A Biblical Guide to Family Planning and Takes

Biblical Map of the Garden of Eden: Where does this Mysterious Garden Exist?

Methuselah: The Biblical Legacy of Noah's Grandfather

Love's Crossroads: The Rewards of Suffering for Love

Features of a Great Christian Camp: A Priority Spiritual Foundation

Daily Mercy: A Journey Through God's Grace Every Morning

Self Help Books by Scott Perdue:

You Are the Masterpiece: Center of the Universe Life Experience

Legacy Blueprint: How to Build a Generational Legacy

Accomplishing Greatness: 10 Legendary Skill Sets of Self-Made Millionaires

The Passive Income Playbook: 10 Game Changing Strategies to Build Wealth

Beginners Guide to Investing in the Future: Gain Wealth from Cutting Edge Sectors

Motivation for Creation: Unlocking the Spark Within

Master Productivity: Unlock your Path to Success

10 Step Productivity Plan: A Guide to Increasing Life's Results

Mindset of Productivity: A Defined Focused Journey

Mindful Love: Embracing Self Love Through Mindfulness and Compassion

Mindfulness for Personal Growth: Transform Your Life One Moment at a Time

The Ultimate Guide to Winning Friends and Influencing People: Master Communication

The Human Connection: Unlocking the Secrets to Understanding and Relating to Others

Stress Free Living: Simple Strategies for Modern Life

Mind Switch: Are you Over-Thinking Negative Thoughts?

Mastering Self-Control: Unleashing the Power of Discipline for Success in Every Aspect of Life

Rising From The Ashes: How to Rebuild When Life Falls Apart

Unlocking Secrets to Weight Loss: A Comprehensive Guide to Science, Nutrition, and Wellness

Effective Diet Supplements for Weight Loss

The Body Detox Blueprint: 10 Essential Steps to Cleanse, Heal, and Revitalize Your Body

Secret 1000 Calorie Cryogenic Diet

Learn to Enjoy Reading: Your Ultimate Guide to Loving Books

The Ultimate Blueprint to Comedy: Your Guide to Mastering Humor and Making People Laugh

Decluttering Your Home: Take Control of Your Space, One Step at a Time

Real Estate Needs Observation: Hot to Bring Light to Entropy & Chaos

Business Books by Scott Perdue:

Legendary Business Skills: How to Think like an Entrepreneur

Seal the Deal: Mastering Sales Objections to Close Every Sale

10 Step Marketing Launch: Ultimate Guide for a Business Advertising Start Up

Email Marketing Success: 10 Ways to Master Business Email Advertising Strategy

Controlled Decent: How to Close a Business

How to Start a Business Networking Group: Learn to Organize and Motivate Business Leaders

Negotiate Like an Auctioneer: Mastering the Art of Persuasion and Control

Auction House Blueprint: How to Win Bids and Host a Successful Auction

How to Run an Antique Shop: Restoring Antique Relics to Modern Living

Secondhand Success: A Complete Guide to Running a Profitable Used Furniture Store

The Thrift Store Playbook: How to Build, Manage, and Thrive in the Resale Business

How to Start an RV Park: Your Roadmap to Success

Turn Rapids to Revenue: How to Run a Profitable River Float Business

Science Books by Scott Perdue:

Creation of Your Galactic Record: Big Bang, DNA, Creation of the Universe. Boom!

Symphony of Life: How Human DNA Plays Like Music

Quantum Cosmos: The Wave Function of the Universe

An Astronauts Heavenly Perspective: Planet, Society and Economy

Earth is the Seed of Life: A Geometric Flower of Life

Infinite Plants in Every Seed

Dodecahedron Earth: Exploring the Geometric Key to the Flower of Life

Pangaea Cracked Open: A Pre-Flood World without Oceans

Ancient Cathedral Architecture: A Language Of Semantics Lost in Time

Power Independence: DIY Guide to Building Off-Grid Energy Systems

Harvesting Heaven: The Ultimate Guide to DIY Rainwater Collection Systems

Farming Tactics for the Sahara Desert: Ultimate Gardening Guide for Arid Takes

Easy to Find Herbal Remedies

How to Build Free Energy Lighting: 10 Effective Easy to Build Free Energy Lights

Dynamic Forces: Exploring the Undeniable Power of Movement

Creative Books by Scott Perdue:

Zeppelin Airship Enterprise: The Future of Flight and Travel Reimagined

Ancient Plasma Energy Weapons Revealed: The Lost Technology of Energy Weapons

Echoes of Camelot: Unveiling the Secrets and Legends of the Knights

Secret Treasures of Rome Revealed: Explore the Ancient Architecture of Rome

The Egyptian Ankh: Secrets of Eternal Life and Ancient Wisdom

Giants, Nephilim, and the Legacy of Humanity: From Ancient Myths to Modern Mysteries

Prophecy of the Seven Suns: Exploring Parhelia in Biblical Prophecy

Epic Scavenger Hunt of Machu Picchu

Adventures of Buying an Island: Edge of Your Seat Suspense Thriller Adventure

My Neighbor is an Inventor: A Journey into Wilson's World of Innovation

Adventures of the Zoo Janitor: Growing Responsibility By Excellence

Exile's Genesis: Chronicles of the New Frontier

Relics of Oklahoma: Route 66 Treasure Hunt

The Oklahoma Waterfall Hunt

www.ingramcontent.com/pod-product-compliance
Lightning Source LLC
Chambersburg PA
CBHW070148230526
45471CB00002B/570